Machines That Learn

A Primer for Teens Who Will Inhabit a World Reordered

Richard Griffith

Russ Hill

Richard Carle

Author

Dr. Richard Griffith, Guilford, VT

Publisher

Dr. Russ Hill, Laguna Beach, CA

Editor

Richard Carle, Medford, MA

Note to the Reader

The vast majority of books are written in the "third person." In those books, "I," "we," and "you" never appear. In this book, "you" crept in to replace references to "the reader." The writing team found this "error" helpful, perhaps because it made the book's message seem more like a letter from a friend. So, we stepped back and started over, embracing that format. The purpose of the book fits the mold of a friendly intervention. Adjusting to life in competition with intelligent machines might prove unhealthy to humans. We want to prepare you for that possibility and arm you with some defenses. That seems friendly. So, we will break the tradition of "third person" writing. But we want to remind you that your English teacher may not allow that freedom when you pen your next school essay. We hope to get away with this intentional blunder because Richard Carle majored in English in college and said, as our editor, we should bend the rule this time.

We left the term artificial intelligence off the cover of this tome mainly to be different. The bookstore shelves are filling up with titles containing the AI label. The concept of machines that can learn forms the central core of artificial intelligence. One of the early pioneers in computer science, Dr. John McCarthy, first suggested "artificial intelligence" as a label for learning machines. He used that term in a conference proposal on computing frontiers at Dartmouth College in New Hampshire in the Summer of 1956. The term "machine learning" appeared in a 1950 paper by Alan Turing. Turing is famous as the creator of the Turing Test proposed to demonstrate computer intelligence. In his test, a human chats via text with two other parties: an unseen computer and a person. If the human cannot correctly identify which responses came from the human, the computer is judged to exhibit intelligence. Today we appear to have computing machines that can pass the Turing test. Similarly, they can defeat humans consistently in various cerebral games, including Chess. So no one can doubt that machines have become intelligent.

I prefer the term Learning Machine, but Artificial Intelligence does seem more popular. "Artificial" can mean fake, imitation, inorganic, synthetic, mock, unnatural, or even insincere. Perhaps man-made represents the best synonym for this use of the word "artificial." The intelligence we tend to believe we will recognize when we see it. We equate intelligence with intellectual capacity, understanding, comprehension, and smartness. We use the same word to refer to information spies might uncover about an enemy during wartime. But in the context of "artificial intelligence" applied to computers, we probably want to stay in the realm of cleverness or problem-solving abilities. Perhaps after reading our book, you will have a better definition of what we call "learning machines."

IBM, an early pioneer in digital computing, has a subsidiary helping businesses and individuals use what it calls AI/LM. They have coined this new designation for their software to link Artificial Intelligence with Learning Machines. They call the company Red Hat because one of the founders, Marc Ewing, always wore his grandfather's red Cornell lacrosse hat while a student at Carnegie Mellon University. Ewing was a star in the computer lab at Carnegie Mellon, and when someone had a question his fellow students would say, "Ask the guy with the Red Hat."

Interestingly, within the community of computer hackers, some are designated "Red Hat Hackers." The Red Hats in this context try to protect innocent folks from malicious computer hackers called Black Hats. The assignment of various color hats to computer hackers appears unrelated to the Red Hat AI/LM software subsidiary of IBM.

Almost a dozen firms have now created machines that converse, so-called Chatbots. Chatbots do precisely the task Alan Turing proposed in the Turing Test. Early in 2023, an innovative group of computer experts calling themselves OpenAI put a computer system dubbed ChatGPT-4 online for the public to try. The OpenAI enterprise started as a non-profit dedicated to ensuring that artificial general intelligence (AGI) systems, which they noted can appear brighter than humans, would benefit all humanity. We know that in the past, wealth created by digital computing increased income inequality in our society. Sam Altman, Elon Musk, and others launched OpenAI in San Francisco in late 2015, hoping AI might have a different relationship with humanity. Their Generative Pre-trained Transformer 4 (GPT-4) chatbot represented an impressive step forward in artificial intelligence that any internet user could experience (www.openai.com) and thus begin to appreciate the implications of the coming era of machines that think. While not perfect, most users found a chat with ChatGPT impressive!

Read this actual exchange (April 4, 2023) between an author of this book and the OpenAI ChatGPT and judge for yourself:

Question: What advice would you give a teenager trying to figure out a career path in the coming age of artificial intelligence?

Answer: "As a language model, my main goal is to assist and provide useful information to those who seek it. Therefore, here are some pieces of advice that I would suggest to a teenager trying to figure out a career path in the coming age of artificial intelligence.

Keep an open mind and be adaptable: The field of artificial intelligence is constantly evolving, so it's important to stay flexible and be willing to learn new skills as they arise. This means that you should keep an open mind when considering career options and be prepared to pivot if necessary.

Learn to code: As AI becomes more prevalent, coding skills are becoming increasingly valuable. Learning to code will give you a strong foundation in computer science, which is essential for building AI systems and working with AI technologies.

Explore different areas of AI: There are many different areas of AI, including machine learning, natural language processing, computer vision, and robotics, to name a few. Exploring these different areas can help you find your niche and determine what interests you the most.

Pursue a STEM education: A solid foundation in science, technology, engineering, and mathematics (STEM) can open many doors in the field of AI. Pursuing a degree in a STEM-related field can provide you with the technical skills you need to excel in the industry.

Look for opportunities to gain hands-on experience: Whether it's through internships, research projects, or personal projects, gaining hands-on experience with AI technologies can help you stand out from other job candidates and give you a better understanding of the field.

Network and seek mentorships: Networking with professionals in the field of AI and seeking mentorships from experienced individuals can provide you with valuable insights and guidance as you navigate your career path.

Remember, the world of AI is constantly changing, and there are many opportunities available for those who are willing to learn and adapt. Good luck!"

The machine's answer contains good advice, but nothing insightful or surprising. We hope to do better in the later chapters of our book. The question we asked tried to explore what careers machines might take away from humans, but ChatGPT wisely avoided such controversy. Later we will revisit that vital issue. We included this conversation with a chatbot to emphasize that thinking machines no longer live in the domain of science fiction. They now inhabit our world and will modify everyone's future.

The Cover of this book lists three authors. Griffith, a retired physician who did graduate work in electrical engineering in the early days of learning machines before attending medical school, wrote the first draft. Subsequent improvements in this book came from Russ Hill, a podiatrist who retired to teach science in the California school system, and Richard Carle, who retired from a career in publishing, mainly textbooks. Three heads are better than one, and we hope you will benefit from this collaboration.

Table of Contents

Chapter 1   So What's the Plan?

Some would say that building machines that can learn and think represents the ultimate accomplishment for mankind. Our book cannot begin to explain all of the technology that this goal requires, indeed we will discuss the fact that important issues remain unknown. Progress toward this goal will certainly create missteps, even dangers, as we harvest the benefits. Teenagers need not care about the technical details, but do need to understand that they will live most of their lives in a world very different from that of all their ancestors because of artificial intelligence. This leap in technical capability will alter the fundamental ways society functions. Imagine an electronic "learning" machine performing difficult tasks more, 1000 times more, rapidly and reliably than we would expect humans to perform.

Then, why should humans struggle to master these same skills? Why spend thousands of dollars for a college degree if a machine has greater capabilities and can work non-stop? How do you begin to chart your life's path in the face of such a bizarre turn of events?   Radical changes are coming, and coming rapidly. Usually, parents and grandparents pass along stories of their experiences that guide the young, suggesting possible paths for success. The actual course of one's life depends upon choices, many irreversible. Making good choices requires focus, curiosity, and diligence, but even then, events commonly send us all in unexpected directions. You will encounter many "forks in the road."

We go to school as a child because society requires it. There you discover the joys of making friends, learning about our world, gaining abilities for self-expression, and learning to think. Our minds accumulate information and competence. We become skilled at organizing and applying knowledge and predicting the consequences of our actions. At the start of schooling, we learn the year that Columbus sailed across the Atlantic Ocean. Soon we are grasping concepts in a more visual context. We discover that our minds have an amazing ability to visualize and manipulate images. We can create pictures, ideas, and relationships that we "see with our mind's eye." You recognize that phrase and understand its meaning even though you do not literally have eyes inside your skull. Indeed, the mind can form pictures of objects that never existed. We can visualize how things we cannot touch might look and function. We can picture relationships. Composers often say they see the music they create as sculptures in their mind. You probably would be surprised to learn that the majority of cells in the

cortex of the human brain devote themselves to processing images, both real and imagined images.

Some would say that we humans think most powerfully when we think in pictures. Probably Einstein had about the most powerful mind of any human who ever lived. Einstein wrote an essay

11

explaining how he discovered new concepts in physics. He said that he sought mental images that might explain experimental observations that had no existing explanation. Although he published his discoveries backed up with mathematical evidence of their validity, the images came first, and often he got help from others to put together the mathematics. Einstein was trying to tell us that human beings appear to think most powerfully using their bizarre ability to create images in our "mind's eye."

The German word "Gedankenexperiment" describes the process Einstein was using to create new insights in physics. Physicists commonly use this term when they explain the rules that govern their science in a visual manner. This term translates into "thought experiment" in English. In relies upon the human mind's ability to form images and to predict the consequences of imagined manipulation of those images. We all do this often, and you may wish to add this word to your vocabulary to help you express the basis of your conclusions to your friends.

If we create mental images of what it means to think and learn, then we can better understand the technology being used to create these exciting new machines we label intelligent. Books commonly have facts, lists, stories, and arguments. But here we want to focus on images, because images allow us to use our most powerful mental capability to see how abstract things, even complex things, work. Once we see something clearly in our mind's eye, we seem to never forget. We can use those mental images in creative ways. In this book we will draw pictures with words and we will sketch other images as actual diagrams or pictures. But we always want to focus on the image, because when you formulate that mental perspective, you truly will understand the how and why.

Einstein searched for the image and then did the math. Let's play with that approach a little. The visual image of arithmetic might be counting pennies. If you arrange pennies neatly into rows and columns you have constructed a visual image of multiplication. The study of algebra might evoke the image of a beam balance or scale. Algebra problems seek to find a number that satisfies specific properties described in an equation. That equation has an equal sign, meaning the symbols on the left "weigh" the same as those on the right side of the equal sign. One uses a letter, often x, to stand for the unknown number you are trying to find. The whole of algebra lies in the premise that if you do exactly the same operation on both sides of the equal mark the equality will continue to hold. Learning algebra is then simply practicing that premise in different situations. That image of a balance scale can carry a student successfully through a couple years of high school math, while students trying to memorize rules for algebra will exhaust themselves and hate every minute.

If you have taken calculus in school, could you visualize calculus? Visualizing it proved essential! Calculus describes problems for which the answer is a function instead of a number. A function describes a relationship between two or more variables. For example, a function can describe how far an object has moved in relationship to the elapsed time. Almost 400 years ago, Isaac Newton, according to the fable, wanted to describe the position of an apple as it fell from a tree. He wanted to quantify the position of that apple at each second of its fall. In other words, he wanted to express the distance it had fallen at 1 second, 2 seconds, and 3 seconds, of even 2.25 seconds. So, we would say he wanted to express the distance as a function of time, and in calculus we would write d = f(t), and we read that as "d equals a function of t" where d stands for distance and t stands for time with f symbolizing the phrase "a function of." What makes calculus fun and powerful lies in its ability to deal with the physical nature of our world in which we have things that speed up and slow down, and we need a form of math to describe position, velocity, and acceleration and other processes that behave in a similar fashion. To do well in the study of calculus, you absolutely need to have a visual image of the process represented by the math symbols. That conversion literally changes the study of calculus from a nightmare into an adventure.

Classes in school, like history and foreign languages, may seem more difficult to make visual. But we still understand and retain information that has a visual component more easily. Please keep that notion of visual understanding in mind as we talk about machines that learn.

Students in their teens begin to study topics that require visual understanding much more than simply memorizing information. These new skills are the ones that prove most important for one's success as an adult. An awareness of this new format in learning will allow you to take full advantage of your mind's great powers.

Chapter 2 What is learning?

Parents ask their children, "What did you learn today at school?" Rarely can anyone express the specific lessons learned at the end of the school day. Nonetheless, we do learn things in school. We know more in the eighth grade than we knew in the seventh. By the twelfth grade, teenagers know pretty much everything, certainly more than parents! But teens lack life experience.

Learning can mean various things. If I learn your name, I have associated a specific word with my memory of your face. The next time I see you, I can say, "Hello, Gertrude." Ideally, you go by Gertrude and not George, and I get kudos for remembering your name correctly.

I have not learned a history lesson unless I can use memorized information to generate consequences and comparisons. If I learn the multiplication table, I have once again memorized lots of associations, this time associations between numbers. If, instead, I learn the content of a history class for an examination, then I remember a collection of events, stories, and perhaps dates that I can regurgitate in response to a question on the test. The teacher wants me to have gone beyond memory to appreciate the importance of specific happenings. The teacher may also want me to compare one situation to another to see if I have understood the consequences of the events I have committed to memory.

My teacher hopes I have learned to think, not just learn to memorize.

If you take a test at school and find that none of the questions seem familiar, you probably did not learn. We now have defined learning in the context of school to include recognizing a failure to learn.

We do lots of learning in places other than school. We taught ourselves to walk before we even developed the ability to remember life experiences. Do you remember learning to walk? No, nor do we know how we did it. Babies want to visit new places and start crawling, even if they never saw another human crawl. Eventually, they decide standing up would be fun, and they work out how to accomplish that process, again without any instruction. Balancing oneself in a standing position has considerable complexity, and you would probably have a tough time explaining to someone how to go about balancing yourself. Your brain carries out that process without making you aware of the necessary details. Humans learn things commonly without being fully aware of how they do so. Interesting!

Learning has an additional element. Humans say they have learned when they improve their performance at a specified task. "Look, I learned how to juggle three balls!" "Listen to this.

I have learned how to play the scale on a trumpet!" "Watch this. I can hit 3-point shots." Probably all learning can be associated with improvement in some performance. Therefore, learning implies a step further than memory, as it contributes to an activity or capability.

Learning goes beyond memory! If we want to understand machines that can learn, we need to identify the elements in the machine's capabilities that allow learning as distinct from memory. We need a clear visual understanding of what we require inside that machine to improve its performance and achieve a specific objective. What goes on inside our brain when it learns? Sorting out these issues has complexity. But we can figure this out!

The machine humans have set about configuring to replicate learning we call a computer. You have most likely used a computer, but you may never have tried to program one or delved deeply into how it works. Teenagers have usually used a calculator also. Calculators do a dazzling job of carrying out tedious arithmetic. Humans invented calculators before computers. The first calculators were mechanical mechanisms starting with the abacus. Next, they had gears that turned and displayed painted numbers to the user to report the desired computational result. Today, we transform those same processes into electronic circuits that turn transistors on and off to carry out arithmetic, often internally using a binary representation of the numbers (only 0's and 1's, no 2,3,4,5,6,7,8,9 allowed).

What can a computer do that a calculator cannot? When you use a calculator, you press keys to input a number. Next, you select a key to indicate the operation you wish to perform and then perhaps enter another number you want to use in the calculation. Computers have the same capability but with the added ability to do something else, something significant that may not leap into your mind.

Perhaps you recognize that the user can program a computer, meaning the user can create a list of tasks for the machine to perform one after another. But programming is not the essential difference. You can purchase a programmable calculator that will perform a long list of calculations one after another, but that does not constitute a computer. We describe the ability that uniquely belongs to a computer as the ability to examine the results of a calculation or operation and then jump to a new place in its list of tasks based on that examination.

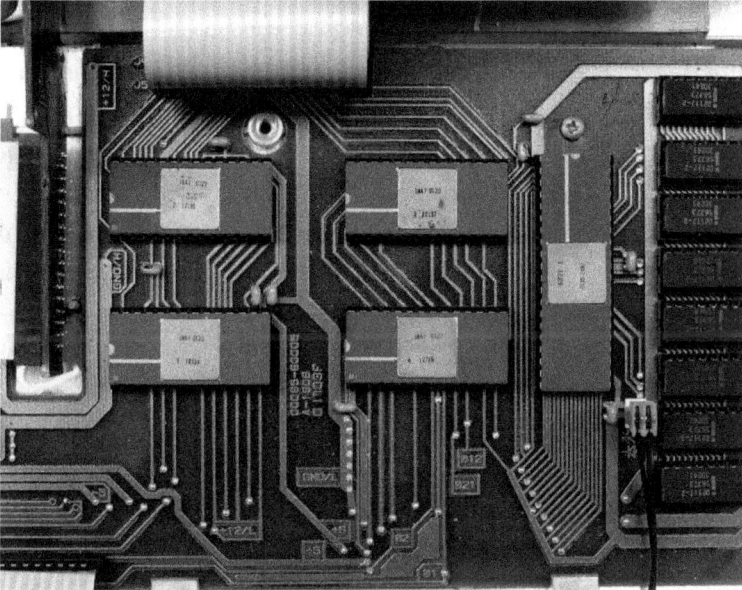

"Examine the results" may need some clarification. The work done inside a computer takes place inside of circuitry dubbed a register. The computer may have more than one register depending upon the size of the computer. The register can do a long list of manipulations on a binary number that it gets from somewhere in the computer's memory. The register can also put its contents back into any location in the computer's memory. When we say that the register can examine a result, we mean that the register can answer questions about the number it is holding. For example, is the number in the register greater than 0? Is the number in the register equal to another specific number stored in the computer's memory? Since a computer really only manipulates numbers, it converts letters of the alphabet into numbers and keeps track of the fact that such numbers are still alphabet letters. But if that particular number equals another number as determined by an inspection, that means the register can be programmed to recognize matching words as well. That capability opens up the opportunity for computers to "read" and "write." Indeed, I am using a computer to write this paragraph taking advantage of that capability. If you think about computers managing the internet, sending emails back and forth for people, and allowing us to shop on line, bank on-line, and look up information about history, sports, or old movies, then you will appreciate that computers spend more time today manipulating the letters of the alphabet than they spend doing arithmetic. But the power of the computer lies in its ability to examine and use that examination to dictate what happens next.

At first glance, that difference may seem trivial, but the ability to jump or branch in the list of tasks opens up a new world of possibilities. Indeed, the power of humans to inspect the results of something they did and then use that knowledge to decide what to do next may prove easier to appreciate as a critical ability. Dogs, cats, and spiders also seem to have this same ability. We are not sure about the amoeba, and no one thinks a germ can decide to change its behavior.

We have discussed learning because of its importance in understanding machines that think. We have learned that the essence of a computer lies in its ability to inspect and use that inspection to alter its following action. But humans do that all the time, and humans believe they have "free will" to decide their actions. Computers do not have "free will" because the program limits the choice of responses. Computers can only act in the manner humans program them to act. But what if we make the program more extensive? Humans have limitations in their abilities. We are not "programmed" with an infinite number of skills. But we certainly are "programmed" to learn, and the intent of this book lies in exploring what sort of programming would allow computers to learn and thus improve their performance on their own. We know that computer learning exists, and spectacular results appear in the news. Some have suggested limits may exist, while the experts are predicting learning machines will surpass human intellectual abilities within decades, requiring only improvements in the programming that makes learning possible. Wow!

In summary, we have noted that learning and memory are different. Learning always involves some degree of memory, but learning goes beyond memory. The evaluation of learning always requires some measure of performance. The story of advances in machines that think has countless references to games ranging from pong to chess. Games offer the advantage of a carefully defined array of options and a measurable determination of performance. Initially, researchers took advantage of the fact that a computer could "play" games rapidly, and thus try out all possible strategies using brute force. But now? Before we leave this discussion of learning, you need to know we have barely scratched the surface of learning. It forms a fundamental aspect of life.

Chapter 3 Let's Do Some Engineering

To develop a robust understanding of the nature of artificial intelligence, we need to talk about both how the brain works and how electronic "brains" function. We flipped a coin and decided to talk about the electronic stuff first. This chapter will start turning you into an electrical engineer. You may feel tempted to think of the designers of computers as "electronic" engineers. But that term has never caught on. Electrical engineers not only study electrical circuits and power transmission, but also have developed a sophisticated understanding of systems able to control machines and processes. That particular area of electrical engineering probably launched the quest for artificial intelligence. It arose explicitly from the need to regulate a process that varied its properties over time. Engineers wanted automatically to adjust the controller to adapt to the changes that might occur in the process.

For clarity, we must step back and introduce you to a few basic, but perhaps unfamiliar ideas. You know that pushing on the accelerator of a moving automobile makes it go faster. Suppose you press the accelerator down a set distance and hold it there. The car will travel faster each second, continuing to increase speed. With the throttle held constant, the velocity keeps growing, but the acceleration remains steady. We call the speed at each second the velocity and the increase in rate with each second acceleration. The units of velocity we term miles per hour. The units of acceleration we call miles per hour squared. In the 1660s, Sir Isaac Newton was trying to understand how these measurements worked together when he invented the field of mathematics we call calculus. By the way, Gottfried Wilhelm Leibniz, a German, created a slightly different approach to this objective of describing processes that store and release energy at the same time as Newton. Today we credit Newton and Leibniz as independent creators of the calculus we rely upon.

To master electrical engineering, you need to feel comfortable with mechanical processes that act like a car does when the driver guns the engine. Mechanical and electrical engineering students, therefore, must study calculus. The same mathematical relationship between position, velocity, and acceleration exists when electricity interacts with components called capacitors and inductors. Like a speeding automobile, these electrical devices store energy. A capacitor consists of two sheets of conducting metal separated by a thin sheet of insulation wrapped often into a cylinder. It can store energy because the electrons squeezed into the capacitor would prefer to spread apart. The squeezing stores the energy. An inductor consists of a coil of insulated wire often wrapped around an iron rod. A current flowing through the coil creates a magnetic field, and the

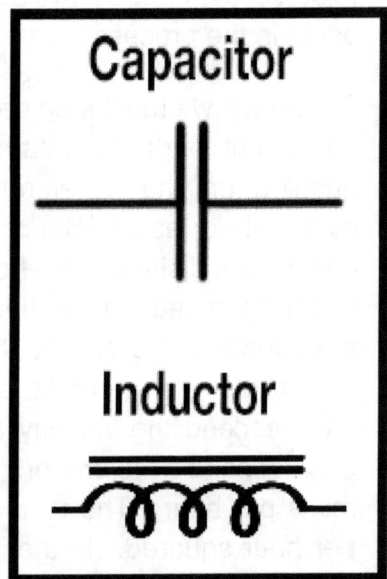

Capacitor

Inductor

magnetic field tries to keep the flow of current from changing either up or down. The magnetic field stores energy that it uses to maintain the flow of current constant. Both of these electrical components (by storing energy) act in a manner similar to the acceleration of a car. Calculating the details of changes in current flow and pressure (voltage) in these components requires calculus because the voltage and current become "functions of time." Engineers study how to use electrical circuits to control various machines and processes. The laws of physics that describe how all of these things interact are crucial for engineers to capture in their mind's eye, including how to express their properties mathematically.

You know that peculiar number in math called pi equal to 3.14159.... with an endless string of digits. That number represents the length of the circumference of a circle divided by its diameter. We think of that number as representing a property of the world in which we live.

There are other such numbers. Fewer people recognize Euler's number, a value equal to 2.71828... also with an endless string of digits. Electrical engineers have a special relationship with Euler's number. Euler, a mathematician who died in 1783, left behind an incredible legacy that electrical engineers use to great advantage every day. Euler recognized that a rare number existed such that a graph of y = e raised to the x power, with e standing in for Euler's number, displays the unique property that each point along its graph has the same value as its own slope. The Nobel Prize-winning physicist, Richard Feynman, called the Euler equation "the most remarkable formula in mathematics." You know what the word slope means. The slope is the tilt of the roof on your house. As every carpenter knows, a pitched roof rises a specific number of feet for each horizontal unit of distance. The roof's rise, divided by that distance, we define as the slope. If you have a curved roof like the Euler curve y = e raised to the x power, the slope has a different steepness at each point.

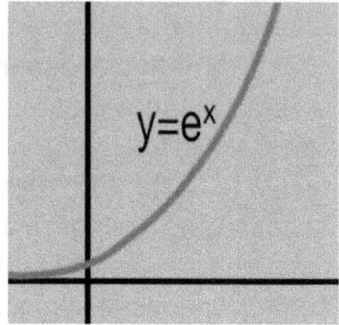

When you look at a drawing of the Euler curve, you may wonder why that feature is so helpful for electrical engineers. The importance does not seem obvious. Math students sometimes take years to capture its significance, if at all. But you can appreciate the idea quickly if you slog through our explanation.

You have a feel for the way an automobile keeps racing along quicker and quicker with a constant push on the accelerator. The Euler equation will describe the unique graph of the car's position if its velocity equals its position at each point along its path. The slope of the graph's curve equals its length along the y-axis at each point x. The value of e raised to the x power always remains greater than zero. The curve passes through 1 when x equals 0. We say the curve grows exponentially (very fast) as x grows larger. Think of the Euler equation as describing the shape of various natural phenomena, especially when energy storage is involved. We see these phenomena in mechanical and electrical experiments involving energy storage in some fashion, for example, loading a spring, filling a capacitor with an electrical charge, or getting a current flowing in a coil of wire to create a magnetic field.

Once you appreciate that Euler's equation has this typical shape, it will make sense that phenomena that display similar behavior will appear as scaled versions of the Euler graphic. For this reason, engineers find it convenient to transform the equations they write to describe circuits and mechanical systems using Euler's work to simplify calculus equations involving slopes into friendlier algebraic formulas. A mathematician named Laplace made this popular. In that way, engineers have distorted the universe of time and space to make their calculations easier. Electrical engineers spend so much time thinking in this transformed space that it starts to feel natural to the extent that they draw conclusions readily without needing to convert their solutions back into the original coordinates of their problem. By the way, have you noticed how important math is?

We have taken this short journey into electrical engineering to give you some appreciation of the tools engineers and other scientists use to understand the world and guide their efforts to create machines to do useful things. A massive piece of human intelligence depends upon an accurate understanding of the laws of nature. Ultimately, learning machines that display intelligence must engage with the same world we live in and be able to make decisions and anticipate how that world will respond in various situations.

We started this chapter by mentioning the need to control processes that change over time. For example, engineers have needed to design systems that can control the flight of a supersonic aircraft traveling faster than the speed of sound. The pilot decides where the plane should travel. But human reflexes would fail trying to adjust the plane's controls quickly enough to keep its flight smooth and stable. The aircraft's performance changes constantly with weather, speed, pitch, and altitude. The center of gravity of the plane shifts as the engines burn fuel. The electronics that control the flight must constantly recalculate the factors influencing the plane's response and plug those numbers back into the control equations to keep the aircraft responding correctly to its environment. Engineers refer to such a control system as an "adaptive" controller.

Do you only find such adaptive controllers on combat aircraft? Not at all. When engineers figure out how to do something fun and exciting, no one can stop them from applying the same lesson in other situations. Traditionally, a home thermostat will turn on the furnace on a cold day and keep it running until the temperature exceeds the desired value. Then the thermostat turns the furnace off completely. Such a system they refer to as a "bang-bang controller." It bangs on when the temperature drops and bangs off again when it comes back up. Today, your home might have a "smart" thermostat with adaptive technology. It measures the amount of heat and time necessary to change the temperature in the building each day and uses that value to plan when it needs to turn the furnace on and off to make the temperature precisely what the owner requests each hour of the day. Such technology may reduce heating costs for large structures, which incidentally helps the environment.

To control various processes, electrical engineers also recognize that they need a feature they refer to as "gain." Gain is the ability to regulate oodles of energy using very little power. Visualize a small boy opening a water spigot at the base of the largest dam. You now have a mental image of gain. The boy may expend ten pounds of force to turn the faucet handle open and later shut, but that action can allow thousands of tons of water to escape. You now have a vivid image of gain.

Imagine a bullhorn. A person speaks into a microphone that changes sound into a weak electrical current. The gain in the amplifier changes the weak current into a much more energetic current flow proportional to the voice pattern. The stronger electrical signal can then drive a speaker cone to vibrate so a large crowd can hear what the person said. Again, you are forming a visual understanding of gain as viewed by electrical engineers.

The construction of a computer requires engineers to use gain. They build electronic circuits that carry out logical operations to process on and off states in large numbers and specific, precise ways. Those circuits store energy, dissipate energy, and exhibit gain. The primary component that creates gain in modern electronics we call a transistor. The transistor makes gain possible through its clever geometric design and materials. A small electrical current can control the ability of a larger current to pass through the transistor. Over time we have learned how to make multiple miniature transistors and other electrical components on a single chip of silicon material so that complex circuits require very little space.

Electrical engineers have calculated theoretical limits on the miniaturization and speed of electronic switching circuits they can create for faster, more powerful computing machines. This calculation "boils" down to analyzing the ability to remove heat from the miniature circuit. A logic circuit in a computer has two or more inputs; based on their status, it decides on its own output. Depending upon the logic at work, that decision destroys or dissipates several input signals. Each signal consists of energy that the circuit must remove as the computer works. Carrying away that energy as heat requires a volume of material, so circuit boards cannot be too small. But if you make computers larger, they cannot work as fast because the circuits must send messages of longer duration since they must travel further. Using this analysis, engineers can calculate the limit of how fast a computer circuit can operate. Over the history of building such devices, engineers have found new ways to construct transistors several times. Such discoveries have allowed circuits to become smaller and faster by reducing their heat production. A leading digital electronics manufacturer recently patented a new geometry for arranging circuit components that may again improve the ability for heat to escape to improve computer speeds and reduce their size.

We will later compare the human brain's circuitry to that of an electronic computer. Now that you have received training in electrical engineering, guess which system, the human brain or a digital computer, has the edge in speed and which in size?

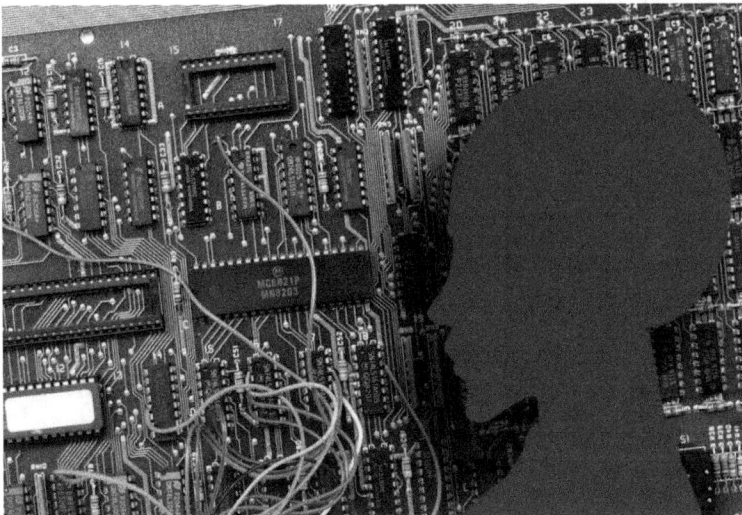

Chapter 4 How Does Our Brain Learn?

The quest to make computing machines more intelligent leads us back to questions about how the human brain works. We know a great deal about the brain, but not nearly enough. Most books on artificial intelligence contain information, often extensive, on the organization of our brain and the functioning of cells that populate that organ. You will learn that the brain's neurons number about eighty-six billion, give or take, and each neuron may have as many as 200,000 connections to other brain cells. You will discover that brain cells operate much like the circuits inside digital computers in terms of each element being at any one time in one of two states, on or off. Nerve cells in the human body fire an action potential or sit quietly as their version of a binary (only two states) operating mode.

Our brain appears to manage particular tasks inside specific anatomical regions. Experts believe language and logic skills commonly live on the left side of our brain, while creative talent resides on the right. We process vision at the back of our head, and motor coordination resides in the lower back of the head just above the neck. Furthermore, our inhibitions live in our forehead. This sort of functional anatomy of the brain's hills and valleys can go on and on, but when finished leaves us with no clue about what goes on inside there when people think!

Still, we have not gotten close to understanding ultimately how we think and learn. Evidence exists that short-term memory in the human resembles some manner of electrical reverberation. On the other hand, long-term memory requires protein synthesis inside cells to make it durable enough to last for years.

It may turn out that we learn more about how humans think by trying to create machines that learn and solve problems than scientists have learned by studying the human brain itself. The fundamental approach to exploring artificial intelligence using computers lies in designing a "backpropagation scheme." Let's define that concept. Backpropagation means the art of using mistakes to improve the thinking skills of a machine. In other words, it means using errors to incrementally fix the performance of a computer trying to learn a specific skill. Do humans learn from their mistakes? We like to think that they do.

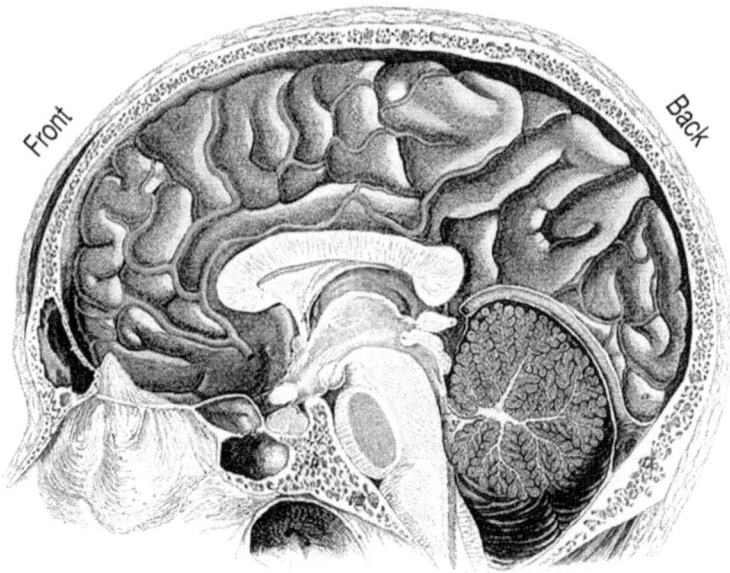

The Post Office processes about 500 million pieces of mail every day. Suppose they asked you to invent a machine that can reliably read handwritten addresses on postal envelopes. Such a machine would replace humans who sort the mail into the bags bound for specific locations. To help you complete this assignment, the Post Office officials gave you a million photographs of handwritten addresses linked to a computer file containing the correctly typed address each handwritten entry denoted. How would you proceed to design such a machine?

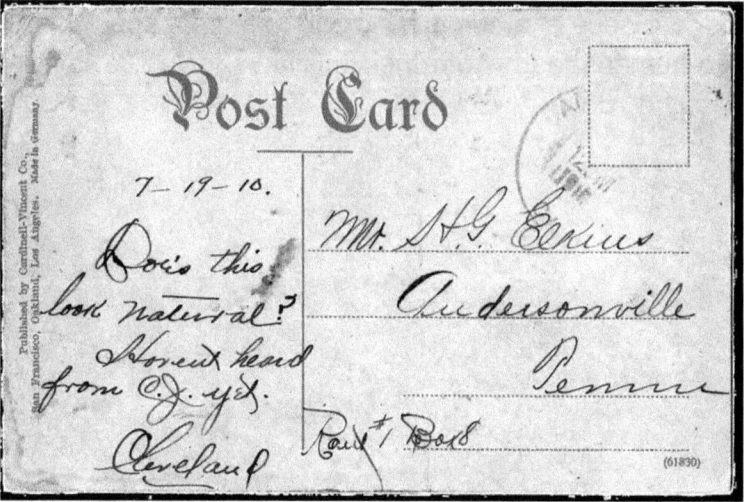

In general, investigators tackling this task have tried to define a specific set of observations they can use to quantify or describe the details of each handwritten address. One might spend a great deal of time trying out different ideas to characterize the various lines in a handwritten address that would allow reliable address decoding. For example, you might start by counting the number of words on the envelope by analyzing the letter spacing. Next, you might try to separate individual letters and begin to make consistent observations about the shape of each letter. Once you figure out a set of features that define each word, you will start formulating a list of characteristics that describe each unique letter. You would like these characteristics to be exclusive for each different character and similar for the same character when repeated in the address or even written by another hand.

The features you select will allow you to begin experiments with your address reading system. One can view each feature evaluated as representing a coordinate (measurement) in a specific dimension. We commonly view the world in three dimensions, but our imagination will allow us to function with more than three. We could view a set of features for a specific letter as defining a point inside a multi-dimensional space. Reading the address would depend upon finding zones in that multi-dimensional area that consistently correspond to a particular letter or character your system is trying to read to figure out the address.

Today, the United States Post Office has machines that read handwritten addresses very proficiently. Engineers developed these machines using an approach like the one we have just described. Originally, they wrote computer programs to methodically adjust the boundaries of the zones for each letter as they found letters that deviated from what they initially expected. Similarly, they often had to figure out new features to make it possible to separate similar appearing letters. Newer versions of such machines may use more contemporary schemes to decode the writing.

How they used the data from the Post Office to guide the development illustrates the role of backpropagation. Each time the procedure resulted in an error, they used that error to move or adjust the boundary of the zone to try to remove that mistake. Backpropagation thus teaches the machine how to avoid making future mistakes. You will undoubtedly appreciate that moving a zone's boundary might fix one mistake but create a different error somewhere else. Does that ever happen with human learning? In learning a new skill, is it possible to disrupt one's ability to perform a task previously mastered?

We have outlined a single way one might use a computer to devise a method for learning how to read handwritten addresses. But if you think about that method more, you will recognize that you did nothing similar when you learned how to read. You did not sit around looking at letters and trying to categorize features that would allow you to determine which one you were seeing. Similarly, you no longer need to spell each word in your mind when you read. Your mind decodes words as a unit even when printed using an unusual font. Our brains do not learn to read like engineers approached that problem for the Post Office.

When humans learn to read, do they use backpropagation? Do you remember reading aloud to someone who stopped and corrected you when you stumbled on a word? It remains unclear how your brain used a mistake to improve its performance the next time you read aloud. That must have happened, but how?

There are about 86 billion neurons in your brain. By now, you realize we are dealing with huge numbers! The nerve cells inside the brain form layers. Each cell has thousands of tiny branches from other cells sending impulses to it, and in turn, when it fires, it sends impulses to many other cells. We believe each cell decides to fire an impulse (called an action potential) when it receives a specific quantity of input impulses from its neighbors. But to learn, the cell must adjust its sensitivity to its inputs according to some consistent recipe. The spread of impulses leaves each cell heading out in all directions. Some appear to go backward, suggesting they carry out the backpropagation role.

Scientists have not yet unraveled many aspects of human brain organization. We know cells in the brain deliver their messages to each other using a variety of chemical transmitters. The axon expels a tiny amount of a specific chemical that activates a protein receptor on the outer membrane of a neighboring neuron. Genetics somehow dictates how cells organize themselves. Yet, we do not understand how that comes about. We know young children can extensively reorganize their brain's connections if an injury or surgery damages a portion of the normal tissue. But as we grow older, that capability fades. Humans seem to learn more quickly in the early decades of their lives than later in life. People who master complex skills must practice those skills for hours, even thousands of hours, if you ask a pianist. The notion that some people are born with highly developed skills never appears to happen as much as we wish it did. But that is another topic.

Humans have evolved sophisticated capabilities to learn. As mentioned before, scientists may learn more about how knowledge comes about than we ever knew before by building learning machines. A machine built by humans but perhaps to some extent designed by a machine will have surpassed humans in its intellectual abilities. Engineers and computer scientists working on artificial intelligence have speculated that they will probably improve their backpropagation techniques several times over the next decade, making them more effective. But then they imagined further improvements in AI would come from the thinking machines themselves coming up with new ideas. If that happens, the rise of machine intelligence will likely resemble the graph of the Euler equation accelerating upward at an ever-increasing speed. And humans will awaken one day to realize that they are no longer the most intelligent entities on this planet.

If you want to learn more about how the human mind processes its surroundings, take a look at **The Illusionist Brain—The Neuroscience of Magic** by Jordi Cami and Luis Martinez. This book goes deep into that topic from a surprising perspective.

Chapter 5 A Closer Look at Neural Networks

The shining examples of learning machines from the recent years of man's efforts to create intelligent machines commonly use an approach referred to as deep learning neural networks. This approach strives to emulate the physiology of the human brain in its essential characteristics. The cells, called neurons, that make up biological brains appear to develop in layers, and neural networks have layers.

Picture a neuron as structured very much like a tree. A region of the trunk would be the neuron's body, with the lower trunk and all the roots representing the neuron's axon. The neuron usually sits quietly until excited. Once excited, it sends an electrochemical discharge down into its roots, eventually connecting with many other neurons. The limbs and twigs of this tree represent the dendrites of the neuron. The dendrites pick up impulses from other neurons and can decide if those inputs, individually,

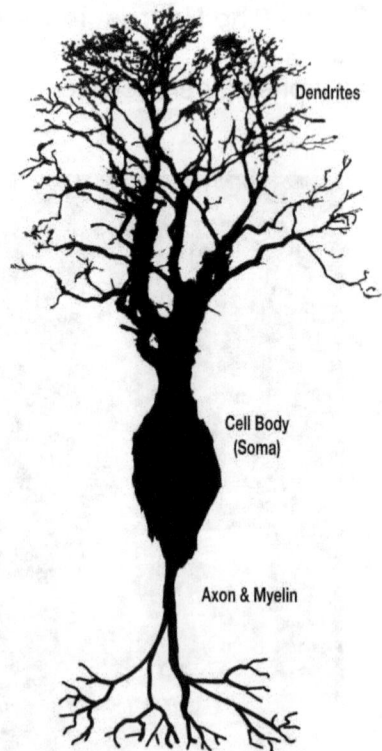

Dendrites

Cell Body (Soma)

Axon & Myelin

should excite or calm the neuron. If the sum of the excitement minus the calming influence reaches a threshold, that neuron fires off an action potential discharge that travels down the axon.

The nervous system has cells outside the brain that carry messages using action potentials. Messages sent to muscles tell them to contract. Messages to various organs modulate their activities. Other nerve cells sense touch, pain, pressure, and other conditions about the body and send that information back to the brain again using action potentials.

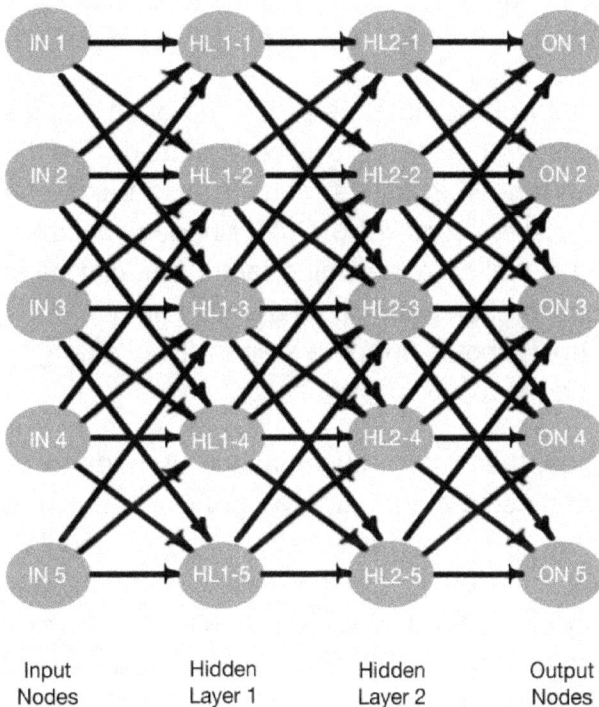

Input Nodes	Hidden Layer 1	Hidden Layer 2	Output Nodes
IN 1	HL 1-1	HL2-1	ON 1
IN 2	HL 1-2	HL2-2	ON 2
IN 3	HL1-3	HL2-3	ON 3
IN 4	HL1-4	HL2-4	ON 4
IN 5	HL1-5	HL2-5	ON 5

We usually think of cells as microscopic, but a single nerve cell in a human can have its cell body inside the backbone or spine and its axon extending down to the big toe. That can make it more than a yard in length. But the cell is still small despite being a yard in length. A myelin sheath may cover such a long axon. You can think of the myelin sheath as akin to the insulation on a wire. The sheath prevents nerve signals from interfering with the impulses of neighboring nerves the same way the insulation inside a telephone cable keeps other telephones using that cable from echoing others' conversations. Some axons lack a myelin sheath, and the action potentials in those nerves do not travel as quickly as those with a sheath. Nerves that control muscles and transmit fine touch commonly need to communicate quickly, so they always have a sheath.

The layout of the neurons inside the brain suggested to scientists that a neural network should also have its nodes arranged in layers. A node acts like a neuron in that it receives signals from other nodes, multiplies each arriving signal by a weighting factor that can be positive or negative, then adds up all of those numbers to decide if that node should fire off a standardized strength signal to the nodes to which it connects.

The idea that many such layers of nodes could be simulated using a computer program and tuned to allow the entire network to do useful functions should not be obvious to you. It was not apparent to the pioneers working on artificial intelligence. Many of them shared their experiences of feeling wonder and awe when their experimental efforts worked, and they successfully "trained" a network to do remarkable things. The term "train" or "training" in a network means adjusting the weights in each node to make the incoming signal more or less critical in the decision process (either more in favor or more against). We next need to consider how one goes about training or teaching the network to deliver the results that will make it worthwhile.

First, let's review for a moment. We have created layers of nodes that resemble the role of a neuron. Each node in the first level (the column on the left) receives inputs from some source of information in the outside world or from somewhere inside the body. Each node uniquely adjusts its sensitivity to each input and remembers that weighting until we decide to change it. Each node sums all of its weighted inputs, and when that sum reaches a pre-determined level, it sends its standard signal (action potential) to other nodes in the next layer up. There may be many intermediate layers called "hidden layers," but then the nodes of the final layer become the output of the entire network. If we designed a network to inspect a chest X-ray in a hospital, the input layer would receive the grey levels of all the pixels in the image. The hidden levels would sort out the presence or absence of various features in the X-ray image. Finally, the last layer (column of nodes on the right) would contain several dozen diagnostic conclusions possible from reading a patient's chest X-ray.

Radiology residents learn to analyze such X-ray images by sitting with a professor who points out the criteria for various conditions one can see in films. Then the physician being trained practices by looking at hundreds of images and comparing their analysis with the results previously recorded for that film. Eventually, the student demonstrates an accuracy sufficient to practice the art/science of diagnostic radiology. We would like to have a neural network do the same.

Training a neural network requires adjusting all the network's weights using images with established correct readings. We need a way of using erroneous network outputs to generate corrections to the weights to improve performance, and we need to keep running this process until the network does not make mistakes. Experts in network training will say that many decisions must come from experience and not theory. The number of nodes and hidden layers required commonly must emerge experimentally rather than by design. Indeed, success often surprises the computer scientists who run the projects as much as it amazes the users.

Board and video games provide a convenient challenge with measurable performance outcomes. A great deal of early work in creating such networks has used playing various recreational games as the laboratory for refining techniques. Stories abound of software designers playing the target game themselves to understand the issues better. But then the machine surprises them when it discovers a strategy that never occurred to the software designers.

The method for adjusting the network weights we call "backpropagation." Computer scientists have written books on the mathematics of backpropagation. Here we want to focus on a visual understanding of the strategy instead of the details of such calculations. Our example of learning to read chest X-rays using hundreds of images previously read by radiologists would be called "supervised learning" in the AI vernacular. We would present each film to our system and use each mistake it makes to adjust our weights until we get the system to work with a high degree of accuracy.

We have drawn a tiny network to illustrate the concept instead of the more complex set of nodes we would expect to require for this complex challenge. Indeed, our conceptual diagram does not contain all the possible connections (arrows) one could add to connect the various layers. We have only put in enough to demonstrate the idea. We could make the diagram more overwhelming by writing on each arrow a value designating the unique weight associated with that arrow. Imagine that the inputs coming into the left side of the network are digits between 0 and 100.

Each arrow would then multiply the value it transports through the network by a weighting value, perhaps between -10 and +10. We have drawn about 57 arrows in our network, but there could be more. In any case, to train this network, we would want to adjust those 57 weights so that the sums accumulated in the output nodes correctly indicate the presence or absence of five specific diagnoses one might make after looking carefully at a chest X-ray. Perhaps ON 1 would be pneumonia, ON 2 would indicate a pleural effusion, ON 3 would suggest a tumor, ON 4 might register a pneumothorax, and ON 5 raise the suspicion of TB. We are trying to visualize the concept and not get lost in the network design details. A serious effort to create a helpful system to read chest X-rays would require a much more elaborate network.

To train our network, we plan to start all our weights at some random setting, meaning the first training scan may not excite any diagnosis. The first scan should turn on node ON 1 because our training set key tells us that the film shows pneumonia. Therefore, our network will make an error by failing to turn on node ON 1. So, we would adjust our weights to learn from that mistake. What weight or weights should we modify, and by how much? That's the critical question!

If you start at the ON 1 node (pneumonia) and trace all the paths back to the IN (Input) nodes, you can tabulate all the weights along each route. That tabulation will allow you to discover which input node influences the ON 1 diagnosis most and which weight should get adjusted the most to get the correct diagnosis. The best recipe for making these adjustments may not be straightforward, and making little adjustments may lead to more stability, slower training, and better performance in the long run. In the classroom, some teaching techniques work better than others, and the same situation appears to be the case for training neural networks.

If you pick up a book on neural networks and look at the discussion of training methods, you will run across the phrase "gradient descent." This term comes from calculus and probably sounds complicated. But if we dig into the concept's visual image, the idea becomes simple and even fun. We are trying to train the neural network so it makes very few mistakes. Think of the weights as spatial coordinates. Our visual experience of this world prepares us to visualize at most three coordinates, up-down, far-near, and left-right. So, let's explore the concept for a three-dimension gradient. We will assume three nodes in our network, then the weights we want to find applied to the range of possible inputs to the network defines a surface in three-dimensional space. Using our training set, we know the desired output for each input. The difference between the output during training and the desired output should define a surface in space. Some call that error surface the "cost" function. The process of training is a process of minimizing the "cost." We can visualize that surface in space as if we were standing on a lawn. And we want to find the lowest spot in the yard because the coordinates of that spot are the weights we seek. The word gradient means the slope of the lawn at each location.

A gradient descent describes how you could find the lowest place on your lawn with your eyes closed. With your eyes shut, you would test the ground around you with your toe in every direction to find the one in which a step would take you down the most. You would take one step and repeat the process. Eventually, you would find a place where a step in every direction would take you up. You would declare that location to constitute the lowest point and, thus, your destination. You would know which weight to adjust and how much. If you are entirely into this visual journey, you might say, "Wait a minute.

Walking step-by-step downward will certainly get me into a hole, but how do I know that I will arrive in the lowest possible hole? There may be a lower valley beyond the one I wandered into, but I would have to walk uphill a little to get there." That astute observation on your part leads to the concept of a "well behaved" surface. Mathematicians define a "well behaved" function or surface as one that does not trick blind walkers trying to find the lowest point. Fortunately, these linear weightings we used in creating a neural network only allow good behavior, so "gradient descent" gets us to the result we want to train a neural network optimally. For a surface to get "badly behaved" requires a much more complicated mathematical construction than you get with simple weights.

Chapter 6 Does My Brain Calculate Gradients?

Oops! Is there a part of your brain that calculates the gradient descent to optimize a student's studies for a mid-term examination? If that capability exists, neurophysiologists have not yet located it. The world would seem easier to understand if the jiffy way we train neural networks were the same process humans evolved to learn new things. But that does not seem to have worked out.

A search of the medical science literature reveals many interesting findings that make one think neurons have some clever process for adjusting the sensitivity of synaptic inputs. Still, it does not appear to involve gradient descent. Scientific publications do conclude that neurons have a system that normalizes their rate of action potential firings over time. We could compare such a system to a voice recorder's automatic volume control. Scientists have discovered important details about how neurons normalize their rate of action potentials. In addition, we know that neurons can increase their connections to other neurons and eliminate rarely active inputs. The details of how they decide such modifications seem less well understood. No one has found evidence that the brain tries to identify the input synapse with the most significant influence to cause an adverse action potential firing and then modifies that synapse.

Let's consider again the case of the youngster figuring out how best to catch fly balls in centerfield so they can patrol centerfield in Yankee Stadium. The details of this learning process remain unproven. Considerable evidence exists

that specific humans are not born with an extraordinary talent of that sort. They may inherit above-average visual acuity or a body suited for a particular sport. However, they must still practice for thousands of hours to master the necessary skills. The best performers correlate well with those who practice the longest and most diligently. Similar observations hold for musicians, artists, dancers, scientists, chefs, and perhaps all other careers. As much as we would prefer it, the evidence does not support blaming our shortcomings on our inherited lineage.

Recognizing intelligence always appears to require measurement of some human dimension of performance suggests that our brain must have a group of neurons somewhere that rewards accomplishment or improvement. Presumably, the performance measure lies uniquely in each individual derived from their singular experience and nature. If that proves true, this center of neurons must feed that evaluation back to reward or diminish synaptic weights. That process would reward hard work and encourage people to put forth the effort needed to achieve their individual, lofty goals.

Evidence from behavioral changes following head trauma would suggest the motivation center of brain function lies in the frontal lobe sitting directly underneath a person's forehead. Individuals with significant trauma to that region commonly have a clinical examination that reveals no damage to their nervous system. Still, their professional careers and social standing seem to veer off course from a lack of enthusiasm. Rugby players do need to wear helmets! Football players suffer similar injuries despite wearing a helmet.

We do not understand how the human brain adjusts the sensitivity of its billions of synaptic junctions to master the many skills a human requires to navigate their life. We know that neurons make such adjustments, and we know much about the chemical processes available to do this job. However, we still do not understand its criteria to guide those modifications. It would undoubtedly prove ironic for an artificial intelligence machine to be the first to discover the methodology of neuronal adjustments that allow humans to reason and learn.

Claude
Shannon

If you have never heard of Claude Shannon, please look him up on Wikipedia. You will be astounded at what he accomplished. You have certainly made heavy use of his genius. Shannon developed a concept of information entropy in the 1940s that allowed information to be measured. The ability to quantify information made it possible to imagine digital communication using binary codes that could eliminate noise that always corrupted (introduced mistakes into) analog communications. We use his concepts when we use the internet, write and save an essay on a computer, or watch a program on high-definition television. Shannon's impact on our lives came from having the ability to measure information, and thus know that a process to encode and transmit that message preserved that measure. He is one of the greats in science. He shared Einstein's view of mathematics as a tool we should use to make science visually understandable. The book **A Mind at Play** by Jimmy Soni and Rob Goodman traces Shannon's life and clearly explains his insights as he forged new approaches to information transmission that we use today in our electronic information systems.

As mentioned before, we demonstrate learning by observing an improvement in some form of performance, but we are just beginning to seek a consistent way to measure learning. A Shannon-like way of measuring learning is evolving. We want to better understand the process by which a network of neurons learns?

The scientific literature describes studies of neuronal physiology that find a network of neurons organizes itself by trying to minimize the error in its expectations of what will happen next. This concept has been called "active inference," defined often as a goal of reducing surprise. This concept differs from the notion that a network learns by maximizing some external reward. This view further suggests that the brain learns by making predictions more accurately or modifying its signals to the body to change the sensations it receives to reduce surprises. Laboratory studies have found evidence of active inference in neurons grown in cell cultures and stimulated by electrical pulses generated by laboratory electrophysiology equipment. Such experiments give weight to the idea that neurons inherently adjust themselves and grow connections or abandon connections trying to minimize surprise and sustain a consistent, moderate activity level. But is that enough of a strategy to create intelligence? Maybe you will be the scientist who answers that question.

We include mention of this yet unanswered question about the human brain to illustrate that research on machines that learn promises to increase our understanding of how humans think. At the same time, this research builds more sophisticated tools to improve the skills of humankind for the future. Remember the term "active inference" because it will probably prove consequential in years to come. And you may be the person to make sense of all of this.

Chapter 8 Can We Make This Visual?

We promised to make sense of this topic visually. And we made the point that the human mind appears to apply a massive amount of its hardware to making sense of images. We all understand cameras and appreciate that a lens focuses an image onto a surface inside the device. Originally that surface was a light-sensitive film that chemically changed to record a photograph. Then we developed the ability to have electronic sensors that recorded the picture as an array of pixels or individual dots that register color and intensity. You may imagine that the retina in your eye is akin to those electronic sensors inside a digital camera such that the optic nerve carries the evaluation of color and brightness from each light-sensing rod or cone back to the brain. However, you would be wrong.

Television initially worked just that way. The early black and white TV cameras sensed the light intensity at each point in the image and transmitted an analog measurement of that value to the receiver in your living room. But TV has evolved to the digital high-definition TVs we use today using a new method of coding developed by engineers at the Sarnoff Laboratory in New Jersey. They discovered they could deliver a more detailed color image to your TV screen by only sending information about changes in the color and intensity of a pixel. If all or parts of the picture did not change, they did not need to retransmit those elements. As a result, they could send less information to the viewers' TVs for their high-definition full color viewing than they had transmitted to the original black-and-white TV receivers.

Next, we talked about computerized neural networks doing this task. They accept many input signals from a scan of an address on a piece of mail and, by a process we call gradient descent, find weighting factors that would allow those signals to only activate a particular node when presented with a specific letter of the alphabet.

When you visualize these two processes, can you begin to see the possibility that they could both achieve the same result of learning to read? In both cases, we have complex calculations that select which combination of inputs meets the criteria for signaling a desired result. In both cases, we have many numbers that we can adjust to get the desired result from a given input. Furthermore, we have a method to make those adjustments by trial and error. We believe the human brain similarly trains itself and has a structure akin to the neural network used in learning machines. We believe the human brain can modify the number of connections/paths during learning and adjust the weights along each path. We usually do not add or subtract paths as we train a computerized neural network, but if we use many hidden layers and nodes, that difference may not have significance. It is customary to connect every node in the machine's network to every node in the next layer up, although we failed to do that in our network diagram to keep the drawing less cluttered.

Your visualization of the neural network learning machine should impress you with the vast number of states it might adopt during training. You can appreciate that training could take many examples and perhaps more than one pass to reach the desired level of accuracy. The speed of a digital computer, in which signals move at the speed of light, allows these neural networks to train more quickly than humans can. The machine also benefits from a memory capability that appears more reliable than human memory. I wish I could remember half as many phone numbers as my mobile phone manages flawlessly.

The individual who edited this book took up the piano after he reached retirement age almost a decade ago. He has practiced several thousands of hours, and you can visualize how the practice has adjusted the weights or sensitivities of his brain's network of neurons. He plays the same melody repeatedly, trying to get each note correct with the right timing and volume. Things he had to concentrate on at the beginning have slowly become automatic (so-called muscle memory, although muscles have no memory), and yet he reports he has a long way to go.

Chapter 9 Randomness and Stubbornness

You did not expect that chapter title! What could randomness and stubbornness have to do with the future of intelligent machines? No one has a very good answer to that question, but you may find it worthwhile to give it some thought.

Engineers have worked hard to make digital computers "deterministic." That word comes from a philosophical doctrine that argues that all events arise from factors external to individual will. In other words, this idea says that free will does not exist. If you are a deterministic being, you cannot just decide to stay in bed all day and ignore your obligations. Humans, especially teenagers, prefer to believe they can make such decisions. But digital computer designers insist that computers have no free will. The machine has to do what the programmer told it to do! If a computer executes a specific program differently today than yesterday, that machine belongs in the repair shop!

Today we have learning machines writing music and creating art. Critics of such activities say, "I don't know about that stuff. It has no heart and soul." They suggest that a deterministic machine cannot replicate the creativity of a human being. This chapter's title proposes that humans seem to introduce an element of randomness and stubbornness that a computer may find impossible to replicate.

Can a deterministic machine act randomly? Computer programmers may find it necessary to create a random number, perhaps to vary the machines' responses to its user to avoid boredom. One trick for doing this uses the clock that main-frame computers have built into their circuitry for administrative purposes. The programmer can access the computer's clock and determine the time of day down to the millisecond. If you want a random integer, you grab the final digit of the time. The probability of capturing any particular digit from 0 to 9 will be about equal. The programmer cannot know which digit the computer will grab ahead of time. So, a deterministic computer can display a degree of spontaneity under these circumstances.

Humans display stubbornness. Despite compelling evidence, they can refuse to accept a specific idea or decision as the ideal choice. Usually, we do not see stubbornness as a virtue, yet in human history, we celebrate some heroes for their stubbornness. They held on to an idea when everyone else thought them wrong. While programming a computer, I cannot think of an instance that I ever wanted the machine to display stubbornness. Indeed, I often wrote a program that the computer failed to execute as I wanted it to. I called that a "bug" and I had to "debug" the program to eliminate that apparent stubbornness. If you have written computer coding, you will appreciate that debugging takes longer than coding.

Many have suggested that we want to create stubbornness as an essential component of all learning machines. We want to create a list of things a machine will never do, regardless of any intent or error by a human programmer or controller. The celebrated science fiction author Isaac Asimov reached this same conclusion in one of his short stories about robots written in 1942:

1. A robot may not injure a human being or, through inaction, allow a human being to come to harm.
2. A robot must obey the orders given to it by human beings except where such orders would conflict with the first law.
3. A robot must protect its existence as long as such protection does not conflict with the previous rules.
4. A robot may not harm humanity or, by inaction, allow society to come to harm.

Interestingly, in 2023, the search engine leader Google published a set of rules it wrote for itself to make artificial intelligence a valued part of its services. Google's intentions match Isaac Asimov's ideas if you grow the notion of physical harm to a broader economic or bias-injury context. If we replace the word robot with one of the many terms that refer to a learning machine, we probably would have the concept of building desirable stubbornness into artificial intelligence taken care of.

Perhaps you can think about this idea further. We have not exhausted the topic of randomness in creativity and the need for stubbornness. Both issues will probably show up in your future in unexpected situations. If you need a topic to write about in an English class, this could dazzle your teacher, especially if you spin it in surprising circumstances.

As learning machines that could adeptly read and write English began to appear in 2023, teachers commonly banned their use by students. A few educators decided that they needed instead to teach their students how to use artificial intelligence efficiently and smartly. As we write this book, the debate on how to do such teaching has just begun.

Chapter 10 Why are You Reading this Book?

Often, we pick up a book to read because the topic seems interesting to learn about. Artificial Intelligence would appear to fit that criterion, but it goes beyond that. Today's teenagers will live in a world changed forever by this technology. By age 20, people generally have formulated critical ideas about what they want to make of their life. Our final path usually bends in unexpected directions and dimensions, but in our teens, we have sorted out many values and ideals that permanently shape us. The power of artificial Intelligence as it develops decade by decade seems destined to impact everyone's life significantly and perhaps unpredictably.

Artificial Intelligence will undoubtedly change the world in a spectrum of significant ways. AI pioneer Max Tegmark at MIT has noted that the automobile's invention dramatically reduced the horse population. The OpenAI ChatGPT Mode-4, in early 2023, allowed the public to ask it questions over the Internet. Many asked it for its view of what AI would do to jobs for humans in the future. It answered such questions in a very modest manner, noting that new technology always brings change, but it usually creates new and different jobs for people. But artificial Intelligence seems a unique new technology because it offers the possibility of performing many current human professions more thoroughly and consistently than mere mortals can ever hope to achieve. And AI works 24 hours daily, with no vacations or sick days.

Humans have held the title of the most intelligent animal on earth for eons, but that distinction may end within your lifetime. Perhaps the length of your life will be increased by machines that learn! Advances in artificial intelligence technology seek to create devices that work faster and more reliably than humans can hope to achieve.

Past generations of humans never had to grapple with an upheaval of their society of this sort. Great Britain created the basic structure of K-12 education for English-speaking students more than a century ago to fill clerical jobs in their government. Perhaps the only significant new courses we have introduced are STEM and driver's ed.

For most generations, young people could plan lives and careers that would be pretty similar to those of their parents, relatives, and the adults in their community. We had models all about us to guide our paths. That seems less accurate at this moment.

Some have looked at artificial Intelligence and worried that it would spawn robots that could take over the world. More realistically, we must fear that nations interested in dominating others could use AI to expand their military powers. Governments understand this and plan to spend billions to prevent this technology from being used to enslave others.

We should not go too far down this road of worry and fear without also recognizing that AI could bring about a new prosperity for human existence in which all humans would not need to work so diligently to maintain a comfortable style of life. Societies could have sufficient productivity levels to allow people much greater freedom of time and resources to enjoy fulfilling vocations and relaxation. Some have suggested that future societies could readily guarantee everyone a high standard of living and support greater equality worldwide. Productivity increases brought about by the development of computer technology appeared in the second half of the 20th century to primarily benefit a small, already affluent segment of society generating higher levels of income inequality. The productivity gains from AI have the potential to dwarf what we saw then, and we should demand that our national leaders should manage this development more wisely over the coming decades than we have in the past.

But what about the decisions individuals must make? A teenager reading this book today has little sway over society but has more control over their future. How could one use their understanding of artificial Intelligence to guide their goal setting and choices of career paths? Answers to such questions, unfortunately, require predicting the future. Usually, we forecast the future by extrapolating from our past. But we have strongly argued that AI makes the past an unreliable model of anyone's future.

Still, some established concepts may prove helpful. Societies that value the welfare of every member appear to offer strength and stability. AI does not seem to change that principle, and we should use that tradition to guide communities in the future. Societies must continue safeguarding the means to feed, house, clothe, defend, police, and provide healthcare. AI has the potential to improve the delivery of all of these services. Humans will want humans to remain in control of all of these roles, but we certainly would not object to machines making all of these services more effective. For example, we would prefer a human to attend to our needs when sick. Still, we would appreciate having machines involved in making medical diagnoses more accurate and treatment methods more precise and reliable. Humans have never figured out how to design enzymes that we might use to manufacture graphene, a material ideal for building houses that would last forever without the need for paint and general maintenance. We would undoubtedly celebrate AI figuring out how to design enzymes, a skill the human mind has yet to accomplish.

Humans enjoy watching other humans perform tasks that take great effort to master. That trait will most likely persist. It is hard to imagine a crowd flocking to see a machine play a musical concert of any genre. Humans will probably always enjoy the arts as executed by other humans. We will enjoy books by human authors. We will enjoy seeing humans act and dance and tell jokes. And we will still celebrate the accomplishments of humans who play sports.

Some would say they would not want to go to court and have a machine decide their fate as judge or jury. Others would counter that they would prefer a machine that knew all the laws without bias to determine their fate. Both positions appear to have merit. You can ponder that decision.

A machine might give me a perfect haircut, but my barber or hair stylist knows me and entertains me with a lively conversation that I would not like to do without.

We now have steel mills that produce steel without human workers, which seems fine. But would I frequent a restaurant where a machine prepared all the food? Would I go to a hardware store without humans on the staff, even if they had a chatbot who would answer all my questions and suggest time-saving solutions to my home repair issues?

Would I want to become an engineer if machines could design bridges, buildings, and devices better and faster than I can? Instead, I might pursue a career in industrial design since that profession focuses on imaginatively analyzing a product or process to find creative solutions that better align with human values, abilities, and preferences.

Probably careers that involve personal relationships will persist in an age of machines having speed and diligence that outshines humans, but charting a course in such a world will have complexity we have difficulty predicting today. We cannot all go into politics, although the future will undoubtedly need talented politicians to preserve human dignity amidst such upheaval. We will probably not elect non-human bots to public office.

The OpenAI ChatGPT Mode 4 predicted that AI would create new roles for humans. Work has traditionally given humans a sense of self-worth and a paycheck. If the number of jobs shrinks and humans cannot find work, we want society to render that deficit harmless. Suppose the increase in productivity from AI means we do not need everyone to have a job. In that case, our society must create new avenues for humans to generate the sense of accomplishment that work previously provided. That may serve as a challenge for you to analyze and hopefully generate creative insights.

With the emergence of AI systems, many leaders in the field quickly called for a new emphasis on the social sciences to lead the planning for various problems they foresee emerging. We immediately need new laws to manage intellectual property rights (patents and copyrights) since AI systems might not care whose ideas they borrow. We need new methods to protect society from fraud and misrepresentation as machines become capable of masterful deception. AI can generate phony but realistic voices and images. No one is saying the machines will become crooks, but a scammer with new AI tools could become especially dangerous and difficult to stop.

This book would prove wonderfully successful if it encouraged readers to start pondering now the challenges AI might best tackle and the new skills humans should cultivate to adapt to a future that will change quickly and dramatically. How can we make sure essential resources are shared? How can we prevent greedy use of the power of these new learning machines so that everyone benefits uniformly? Has any generation ever faced such enormous challenges emerging so quickly? Perhaps not.

Chapter 11 The Traits of AI

In the last chapter we mentioned that AI has the potential to make human life more enjoyable for humans, but like any new capability it can lead to good or evil. Science fiction writers have explored this topic for decades, especially the evil side of the coin, because good rarely makes for an exciting story. But now we must all begin dealing with such issues in our day-to-day lives.

In the science fiction arena, we had to worry about robots taking over the world using their super-human speed and accuracy to dominate and enslave mere mortals, or simply wipe them from existence totally. Now we appear to feel more confident that as long as AI needs electrical energy to function, humans can pull the plug if necessary.

In the meantime, we have already learned that early chatbot systems, while impressive in their ability to put together concepts quickly, still exhibit errors and hallucinations as they put together ideas without benefit of the commonsense humans acquire from worldly experiences. Fixes will certainly emerge, but that may mean errors will simply become more elusive.

Many have asked if AI systems will become self-aware, sentient, prescient, conscious. The definition of these words may not be familiar. You may feel comfortable with the term self-aware, which generally means having knowledge of one's own feelings and character. Sentient refers more to the capability to perceive or feel. And, prescient refers to the ability to anticipate the outcome of actions or events. Conscious may represent the easiest target since it only means the ability to respond to one's surroundings. Many scholars have stated with great certainty that machines can never become self-aware simply because they are machines. That seems a lazy philosophical conclusion.

Humans have long pondered the degree of self-awareness exhibited by various animals. Interestingly, some have suggested a test of self-awareness for one's pet dog. The test calls for one surreptitiously to place a small piece of tape on your dog's head outside of its field of view or awareness. Next, you allow your pet to view itself in a mirror. If the pet attempts to paw the tape, according to this test, you have demonstrated self-awareness. Your dog may not pass this test. Experts report that chimpanzees, dolphins, and elephants reliably recognize themselves in a mirror, along with the bird we call a magpie. The magpie is a member of the crow family. Crows have been reported to make tools on occasion to help them get at food. Perhaps being called a "bird brain" may not constitute an insult.

You may object to the notion that recognizing oneself in a mirror represents self-awareness. Interestingly, designers of self-driving cars have found a need to consider a similar issue. If a car approaches a reflective store front, will it recognize itself or instead perceive that another car is driving directly toward it?

The issue of self-awareness in AI systems should ideally challenge us on a deeper level. The question provides an opportunity to think more deeply about what it means to be human. We have a biological computer with billions of neurons that learns both independently from experience with its environment and from instruction. It experiences pain from an array of causes. It can act on its surroundings and evaluate the consequences of those actions. All of this contributes to an awareness that includes an understanding of the possibly of its own death. That awareness also includes the trait of empathy, which means our brain recognizes that others think and feel similar aspects of awareness.

The complexity of human awareness would certainly encourage one to conclude that a machine cannot reproduce that quality. The opposite conclusion might arise from noting that our mind derives its capabilities from the capabilities of billions of neurons. We believe that all the things a neuron can do could also be designed into an electronic circuit or simulated by a digital computer. If that is the case, we could say we know of no reason a machine could not eventually display awareness equivalent to human awareness.

Now you can write a science fiction tale in which you learn how to transfer all of your experiences and knowledge into a digital memory bank for an AI computer to use to become you. Of course, in that manner you have become immortal. You will, of course, need to work out which of the two forms of you is the real you.

Chapter 12 Learning Machines and the Law

Public exposure to Chatbots on the internet in 2023 resulted in a considerable debate in the media about the safety of machines that think. Indeed, some prominent public figures suggested society should declare a moratorium on artificial intelligence research and development.

We all recognize that almost anything that has the power to change society can have negative and positive consequences. This sudden reaction against AI starkly contrasts with the Executive Order of the President of the United States on February 14, 2019, entitled Maintaining American Leadership in Artificial Intelligence. That document stated, "Artificial Intelligence (AI) promises to drive the growth of the United States economy, enhance our economic and national security, and improve our quality of life." The executive order references the need to maintain "safety, security, privacy, public trust, and confidentiality protections."

Do we need a moratorium? What is it that frightened Americans? Some read remarks by Oxford University Michael Osborne, a professor of machine learning. Osborne suggested that AI could eliminate humanity when it becomes more intelligent than humans. Others have related the concern that AI used for military purposes could make nations more capable of destroying their adversaries.

On a smaller scale, many have proposed that AI used to synthesize sophisticated fake reports, photos, and videos could lead to public hysteria and uprising. That concern has historical evidence to support it as a real issue. Lies have promoted violence, and if one can convert a lie to realistic photographic or video evidence, it can indeed represent a powerful tool to excite people to action. CGI (Computer-Generated Imagery) has allowed moviegoers to witness imaginary events that have seemed natural for years. AI can make such a capability available to anyone who wishes to confuse a neighbor, a crowd, or a social media following.

You would probably agree that society needs laws to prevent using AI or other means to scam others. But it also seems problematic that laws to prevent such scams may no longer be adept enough to stop damage or injury if individuals can no longer distinguish the truth from falsehood. That concern has prompted the notion of preventing the development of these AI tools.

To stop all AI development reminds us of the adage, "Don't throw the baby out with the bath water." No one wants to stop the advancement of learning machines that can improve science, streamline our legal system, and make our economy more productive. No one wants to prevent the development of learning machines that can make medical diagnoses more accurately to potentially save perhaps half a million lives each year around the world. We instead hope that humans can find a way to use the productivity of AI to make humans more secure, healthy, and joyful in their lives. That would seem to mean we must modify laws.

Chapter 13 Preparing for a Life with AI/LM

The authors of this book have lived long enough to remember life before the internet and cellphones. Those inventions changed their lives. We wrote this primer on learning machines especially for teenage readers. You will live and work most of your life in a world changed dramatically by the advances made with machines. These machines that think will work relentlessly at dazzling speeds. Machines will drive cars, will figure out your tax returns if that still exists, will work out your medical diagnoses, sooner and more accurately than human practitioners can to keep you healthy. Humans commonly take pride in their careers using work as a measure of self-worth. That may change.

The prospects seem exciting, yet perhaps frightening! Today we think of careers lasting approximately four decades. Life expectancy will increase in the coming decades, so people may want to retire from the workplace later. But regardless, forty years' worth of advances in the capabilities of learning machines defies our ability to predict. We expect learning machines to contribute to their own improvements within that period, so design progress should accelerate. Perhaps the adage, "You ain't seen nothing yet," applies.

You are accustomed to homework. Homework usually involves writing, reading, memorizing, and solving math problems for practice. Machines will probably change those activities as they play a more active role in the education of humans. We already have evidence that computers can improve human schooling by facilitating information retrieval practice. We hope this will help students learn faster and more effectively.

How will learning machines change careers and roles in society? When you next have to write a school essay, perhaps you might think about how AI could make your life better or different. Most importantly, it would help if you thought about preparing yourself for a life that will differ markedly from previous generations.

Your essential task will involve adjusting your personal goals and objectives. A century ago, a high school education seemed sufficient for success in life. A generation ago, a college degree seemed necessary. With the arrival of machines that think, perhaps humans will need a doctoral-level education.

Humans will probably continue to value anyone who excels at sports and the arts. If you decide the safest path for your future lies in becoming a professional athlete, you will still find machines that think will alter your world. If you elect to make your fortune playing the saxophone, a machine may write your music and polish your recordings.

Today we have thousands of jobs employing humans in military roles. Replacing them with intelligent robotic talent has clear benefits for humankind. Jobs in manufacturing have been shrinking for years, but may soon totally disappear. Construction careers may persist, but some experts have suggested more construction will move into a factory environment in the coming decades with much less human labor.

Perhaps you will elect a life as a hermit. That could work. Society could hopefully harness the productivity of learning machines to create a system that makes the necessities of human life freely available to one and all. Making that happen will require talented public officials, and perhaps you should carefully consider a future in politics. If society does decide to make life essentials universally available, you could devote your energy to fiction writing or oil painting without worrying about how to pay the bills.

These are all things you need to ponder as your homework. They can seem strange concepts, but rank as crucial issues for your generation. Make a list of possible careers or challenges that interest you. See if you can predict which of them might prove most resistant to being taken over by machines that learn. Then ask yourself

what sort of education you would need to excel in those endeavors.

Will making such a list make a difference? Life coaches would argue that it does. They would add that people commonly find unexpected events change the path of their lives, taking them entirely away from goals they may have set initially. However, they would also note that focusing on those early goals took them down paths they needed to travel to achieve success in their final destination. Setting goals proves necessary even when those goals later need remodeling.

Exciting times lie ahead in your life, with more changes and surprises than anyone can imagine today. Popular Mechanics Magazine, in its May/June 2023 issue, looked at ChatGPT with Sameer Singh, an associate professor of computer science within the University of California system, and concluded, "Will it put us out of work? I think some people may be thinking that, but they just need to play around with it for 10 minutes, Singh says, seeming unfazed. 'It's not happening.' He summed up ChatGPT as a 'very sophisticated guessing engine,' and we think that says it all."

I would encourage you not to take advice from Sameer Singh and Popular Mechanics. The baby version of ChatGPT has many faults, but the coming generations of ChatGPT will arrive much more rapidly than human generations. They will evolve talents far more quickly than human evolution. Within your lifetime, machines that learn will put humans out of many current human careers.

Thank you for reading our effort to provide a visual appreciation of machines that learn. If it encourages you to set new goals or better understand the news of recent achievements in artificial intelligence, this will prove a worthwhile endeavor. You can play an essential role in making the new human society using this technology, one that fosters new dimensions in civility, nurturing, and trust.

Comments from the Authors

GRIFFITH: Some of the happiest moments of my life I spent dealing with computers. As a teen, I took a soldering iron in hand to build an electronic calculator in the late 1950s as a home project suggested by Popular Electronics Magazine. As a physics major in college, I wrote programs using the University's first digital computer in a language called Algol. Then as a graduate student in electrical engineering at Rensselaer Polytechnic Institute, I used an IBM 360 Serial #1 programming in Fortran. RPI also had a more petite General Electric computer that I used to practice Assembler language programming. After earning a doctorate in engineering, I entered medical school in Richmond, Virginia. Their neurosurgeons crafted a multi-discipline research program to improve head trauma care. They needed computer support to manage clinical data, monitor evoked potentials, measure cerebral blood flow, and track intracranial pressure. I worked on their research grant after finishing medical school for two years before doing a residency in anesthesiology.

Later I worked with a talented MIT computer scientist to develop a "crowdsourced" system to tract gems of wisdom for the delivery of anesthetic care tied to specific surgical procedures as done by individual surgeons at the Albany Medical Center. Sadly, we could never fund the full implementation of that tool designed to augment human memory and reduce errors in care.

I have previously written two books with Dr. Russ Hill. This book on the hot 2023 artificial intelligence topic will make some critical issues more available to bright young minds.

Richard Griffith

HILL: I practiced medicine for almost 20 years, and following retirement taught environmental science, pre-medical science, and pre-engineering. I have published youth-focused books on medical careers, analytical thinking, and environmental impacts on health. I co-authored "Medical Investigation 101" with Dr. Richard Griffith and "Why Did Buffy's Fur Go Flat?" with my daughter, Erin Hill.

My current interest in advanced water technologies grew from the waste processing and water reuse challenges I witnessed in my community. After investing in advanced solid waste technology, I spent several years studying this science before joining BrightWater HDC Inc. as Co-founder and Vice President. I feel honored to have worked on the "Machines That Learn" writing team because I firmly believe this advance in technology will become life changing for today's youth and tomorrow's society at large.

Russ Hill

CARLE: After a career in educational publishing, I retired just as that industry began evolving into the digital age. I admit to spending way too much time now playing online bridge and crossword puzzles, trying to play scales on a piano perfectly, and pursuing an odd interest in the history of the Pennsylvania Railroad. I have no idea what Tik Tok is.

I was a classmate of Richard Griffith's at the University of Virginia shortly after Thomas Jefferson founded the school. I think I qualify as a liberal artist, and they needed me to make sure this primer on a complicated technology would make sense to real people.

Richard Carle

Suggested Reading List for Those Aspiring to a Better Understanding of Machines That Learn

1. *Artificial Intelligence For Dummies* (2nd Edition) by John Mueller and Luca Massaron

2. *Python: Beginner's Guide to Artificial Intelligence* by Denis Rothman, Matthew Lamons, and Rahul Kumar

3. *The Hundred-Page Machine Learning Book* by Andriy Burkov

4. *Life 3.0 Being Human in the Age of Artificial Intelligence* by Max Tegmark

5. *Artificial Intelligence Basics: A Non-Technical Introduction* by Tom Taulli

6. *Machine Learning for Absolute Beginners: A Plain English Introduction* by Oliver Theobald

7. *A Human Algorithm: How Artificial Intelligence is Redefining Who We Are* by Flynn Coleman

8. *Machine Learning for Dummies* by Paul C. Mueller

9. *Artificial Intelligence and Machine Learning* by Jason Brownlee

10. *Deep Learning* by Ian Goodfellow, Yoshua Bengio and Aaron Courville

11. *Machine Learning for Hackers* by Drew Conway and John Myles White

12. *Machine Learning for Humans* by Vishal Maini and Samer Sabri

www.ingramcontent.com/pod-product-compliance
Lightning Source LLC
La Vergne TN
LVHW051814080426
835513LV00017B/1955